by Elizabeth J. Natelson

Harcourt
SCHOOL PUBLISHERS

Orlando Austin New York San Diego Toronto London

Visit *The Learning Site!*
www.harcourtschool.com

Introduction

When you help someone make a cake, you probably use a measuring cup to put the right amounts of flour and sugar into your bowl. You use other measuring tools around the house, too. For example, what would you use if you were sick and thought you had a fever? You would use a thermometer to check your temperature. How would you measure the distance from one side of your bedroom to the other side? You might use a yardstick or a tape measure.

But, how would you measure the distance from Earth to the planet Venus? The scientists who work at the National Aeronautics and Space Administration (NASA) have that answer.

NASA has sent spacecraft to explore most of the planets in our solar system. In addition to determining the distances between planets, the spacecraft explore the size and shape of the planets, what their surfaces are made of, the temperature of the planets, and what gases exist in their atmospheres. Each spacecraft carries an amazing set of instruments specially designed for one purpose: to measure a planet.

Seismographs Measure Movement

Scientists measure Earth more than any other planet. The plates that make up Earth's crust are always in motion, so that makes Earth an interesting and complex planet to measure. Every year, Earth has about one million earthquakes that are so small that no one can feel them. In that same time, there are many more earthquakes that have an intensity strong enough to be felt or to do damage.

Each earthquake sends a seismic wave through the ground. *Seismic* comes from the Greek word meaning "shake." Scientists measure these vibrations with seismographs. *Seismographs* tell scientists where an earthquake is, how big it is, and what kind of Earth movement caused it. The seismograph information from an earthquake can help scientists learn why the earthquake happened.

Seismographs can measure where this earthquake is and how strong it is.

Scientists also use seismograph information to measure how earthquake waves move through the ground. This helps them understand about Earth's layers: the inner core, outer core, mantle, and crust.

Scientists can also use seismographs to gather information about the moon. Astronauts from the Apollo Project set up six seismographs while they were on the moon. These seismographs have detected from 600 to 3000 small "moonquakes" every year. Many of these quakes were found to be caused by meteors crashing into the moon's surface.

By measuring the movements of moonquakes, scientists have learned that the moon's crust is 70 kilometers (45 miles) thick in some places. Seismograph information has helped scientists learn more about how the moon may have formed.

Sonar Measures the Ocean Floor

Earthquakes and even volcanoes can be caused by moving plates that make up Earth's crust. These moving plates can be observed on the ocean floor by using sonar echoes. Scientists use a sonar (SOund Navigation And Ranging) instrument to study Earth's surface under water.

Light does not travel well under water. Cameras can take good pictures only to about 10 meters (33 feet) deep. Sound waves travel better through water than through air, so scientists invented sonar.

A ship can send a sound signal into the water. Then sonar measures how long it takes for the echo of the sound to bounce back to the ship. The deeper the sound signal has to go before hitting something and bouncing back, the longer it will take for the echo to return. By measuring sonar echoes, scientists have mapped much of the ocean floor. Where they once believed the ocean floor to be flat, they have found high mountain ranges, called ridges, and low trenches.

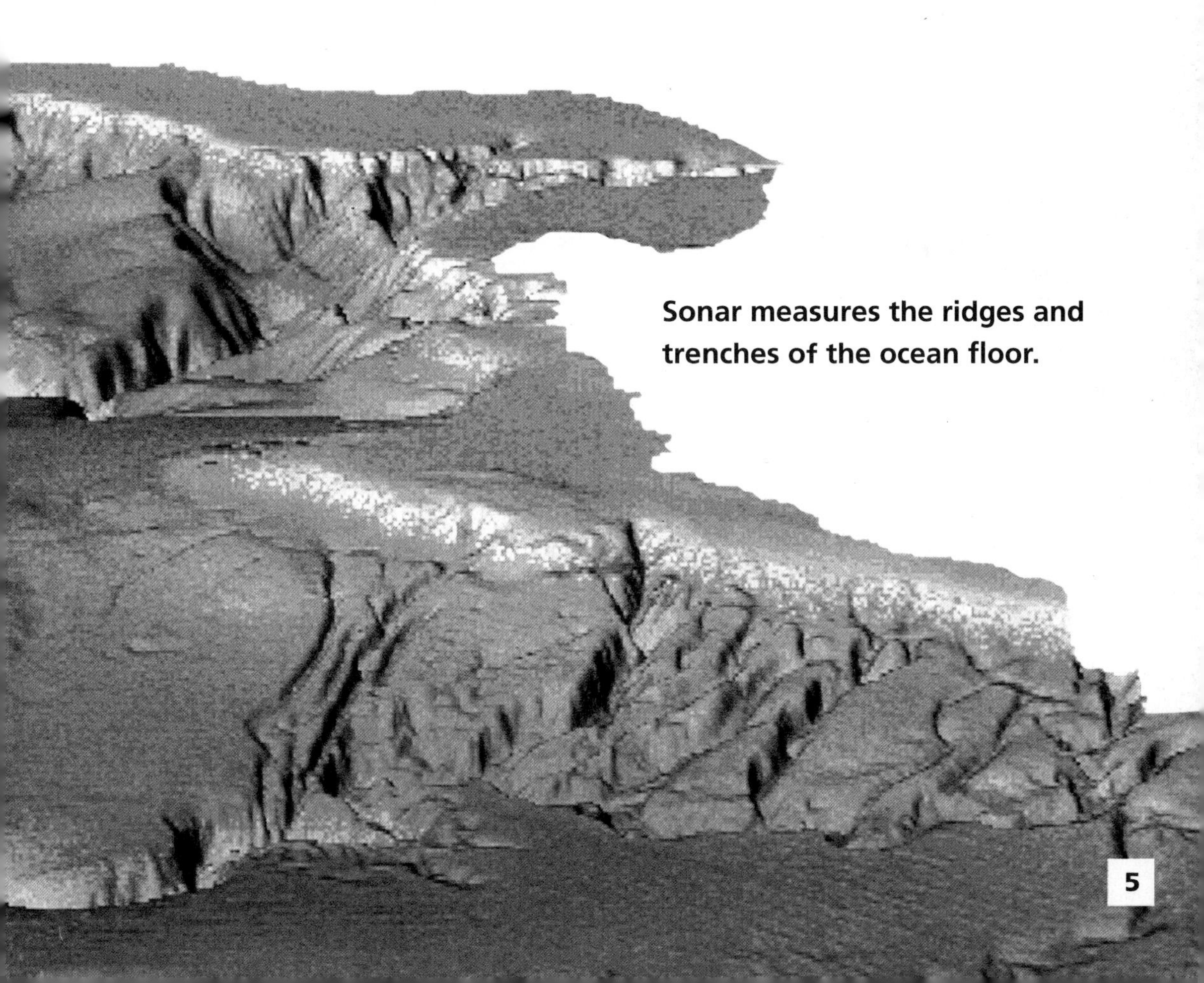

Sonar measures the ridges and trenches of the ocean floor.

Europa may have an ocean beneath its icy crust.

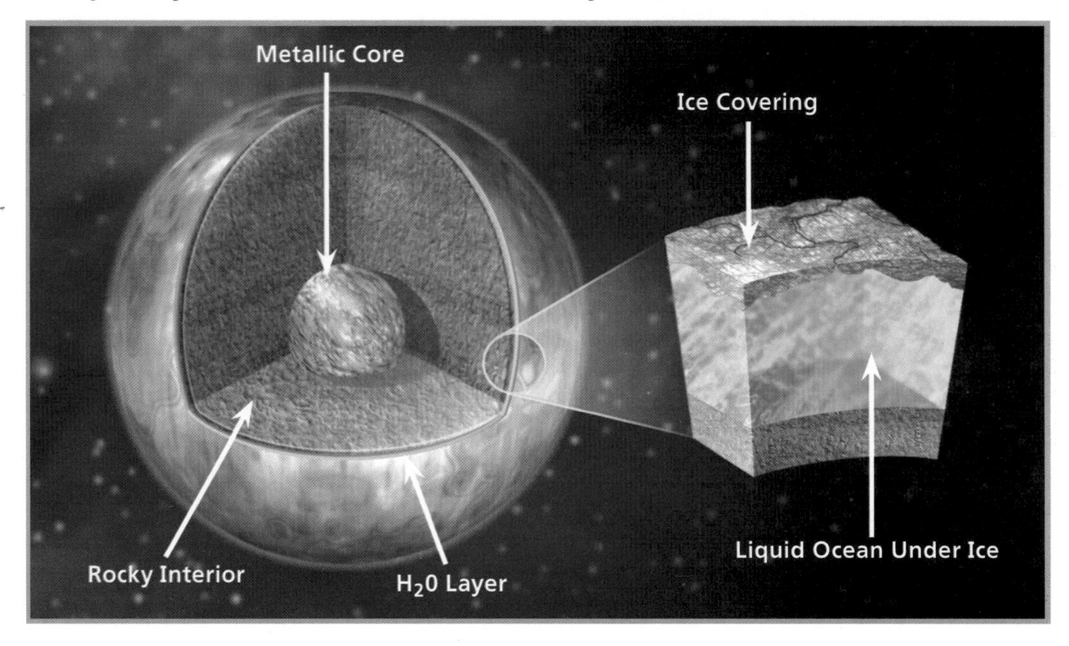

Modern sonar systems can also measure how strongly an echo returns. The strength of the returning echo tells scientists whether the sound has bounced off solid rock or soft mud.

So far, sonar has not been used to measure other planets or their moons. This is because no oceans are known to exist anywhere in our solar system other than on Earth.

However, Europa, one of Jupiter's moons, might have an ocean underneath its icy crust. Scientists have learned much about Europa from the *Galileo* spacecraft. This spacecraft studied Jupiter and its moons for eight years, from 1995 to 2003.

The fourth largest of Jupiter's moons, Europa is about nine-tenths the size of Earth's moon. A very thin atmosphere of oxygen covers Europa's surface. The surface itself is smooth and covered with a crust of ice only 5 kilometers (3 miles) thick. Scientists suspect that beneath that ice lies an ocean that may be 50 kilometers (30 miles) deep. Heat from inside Europa might be keeping that ocean liquid. If there is an ocean under Europa's ice, sonar may be used to map its floor someday.

Radar Through Darkness and Clouds

Just as sonar sends out sound waves through water, radar (RAdio Detection And Ranging) sends out electromagnetic waves, called signals, through the air. When the signals are sent out, the radar measures the time it takes for them to bounce back, as well as the strength of the returning signal. The radar operator can then determine how far away an object is. A radar signal can go through clouds or darkness as easily as sunlight can go through a glass window.

There are many different kinds of radar. Some are set up on Earth's surface, some are carried by airplanes, and many orbit Earth on satellites.

A radar altimeter on an orbiting satellite reflects a signal off the ocean's surface. From the return signal, or echo, the radar altimeter can measure the satellite's height, or altitude, above sea level. This data is used to study how the sea level may vary over time. It can also be used to monitor the water levels in lakes and reservoirs.

Other satellites use SAR (Synthetic Aperture Radar) to send signals to Earth. From these signals, SAR can measure the echoes to see how quickly they are returned and how strong they are.

Radar from an orbiting satellite made this image of a volcano in Russia.

Scientists can use SAR to learn, from orbit, where plants grow on Earth's surface and how high they grow. They can also use SAR to see how things have changed as volcanoes, earthquakes, and glaciers continually alter Earth's surface. Computer images of Earth's surface are used to study these changes.

For example, an earthquake can change the shape of the land by causing the land to split or crack. This change may be large or small. Using SAR, scientists can measure exactly how much the ground moved. They can also measure changes in Earth's surface that may indicate that an earthquake may soon happen or that a volcano may soon erupt. SAR is very sensitive and can detect very small changes. From Earth orbit, satellites can measure and map centimeters of land movement caused by earthquakes or even movements of Earth's crust that would otherwise go unnoticed.

SAR was also used to map the surface of Venus. Venus is an extremely hot planet with a surface temperature of 460°C (860°F). Its atmosphere is thick and poisonous. No astronauts could ever visit there, nor can instruments sent to the surface last very long in the melting heat and corrosive atmosphere. A spacecraft orbiting Venus cannot "see" through the thick, swirling clouds. The best alternative is to use SAR.

Radar signals can pass easily through clouds. From 1990 to 1994, the *Magellan* spacecraft used SAR to look through the clouds and map almost all of the surface of Venus. *Magellan* found mountains, valleys, and plains on Venus. *Magellan* also found that Venus is covered everywhere with volcanoes.

Radar can "see" through the clouds of Venus.

Venus with radar "eye"

Venus with naked eye

The *Cassini* spacecraft has a similar job in a different part of the solar system. One of its tasks is to use radar to map the surface of Titan, Saturn's largest moon. Of all the moons in the solar system, only Titan is known to have an atmosphere.

In October 2004, *Cassini* used its radar to take pictures of Titan's surface. The pictures showed dark areas that might be lakes of organic materials called ethane and methane. Those materials are gases on Earth, but on Titan's frigid surface they flow like water. Radar also revealed shapes that look like ridges of sediments piled up by wind. *Cassini* also saw volcanoes. Instead of erupting lava, however, Titan's volcanoes most likely erupt water and ice.

Laser Uses Tightly Aimed Light

A laser sends light toward an object, just as sonar sends sound waves and radar sends electromagnetic waves. A laser (Light Amplification by Stimulated Emission of Radiation) produces a tight, organized beam of light. Light from a laser is one color and is aimed in only one direction instead of spreading out the way light from a flashlight does. Scientists use special laser instruments to study Earth and other planets.

One special instrument that uses a laser is called LIBS (Laser-Induced Breakdown Spectroscopy). A LIBS device, which is only the size of a flashlight, shoots a strong laser flash at a sample of soil. The powerful beam turns a tiny bit of that soil from solid to gas. When that happens, the LIBS detects how much carbon is in the soil.

Another special instrument is lidar (LIght Detection And Ranging). Lidar sends a beam through a laser. It measures how light reflects back from gases in the air. Using this instrument, scientists study clouds and other parts of the atmosphere.

A third instrument that uses lasers to study planets is the laser altimeter. This device measures the distance from the altimeter to the ground, much as a radar altimeter does. A laser altimeter sends the signal from a satellite down toward the ground. It measures the time it takes for the signal to come back. If the signal bounces off a mountain, the signal will return sooner than if it goes down into a valley. A laser altimeter is used to map a planet's landforms.

A laser altimeter is one of the instruments aboard the *Mars Global Surveyor.* This spacecraft began its orbit around Mars in 1997. The job of the laser altimeter on board the *Surveyor* is to map Mars's surface and to measure the height of the clouds.

While analyzing Mars's surface, the laser altimeter gathers one bit of information for each pulse of light it sends and receives. The altimeter gathered 2.6 million pulses to put together a computerized surface map of Mars's north pole.

Another laser altimeter is also aboard the spacecraft *Messenger.* This spacecraft took off for Mercury in August 2004. The laser altimeter was sent to learn about the hills, cracks, and craters of Mercury's surface.

Spectrometers

Sonar, radar, and laser instruments have something in common—all of them are active. *Active* means that these instruments send out a signal. These signals include a sound wave from sonar instruments, electromagnetic waves from radar instruments, and a beam of light from laser instruments.

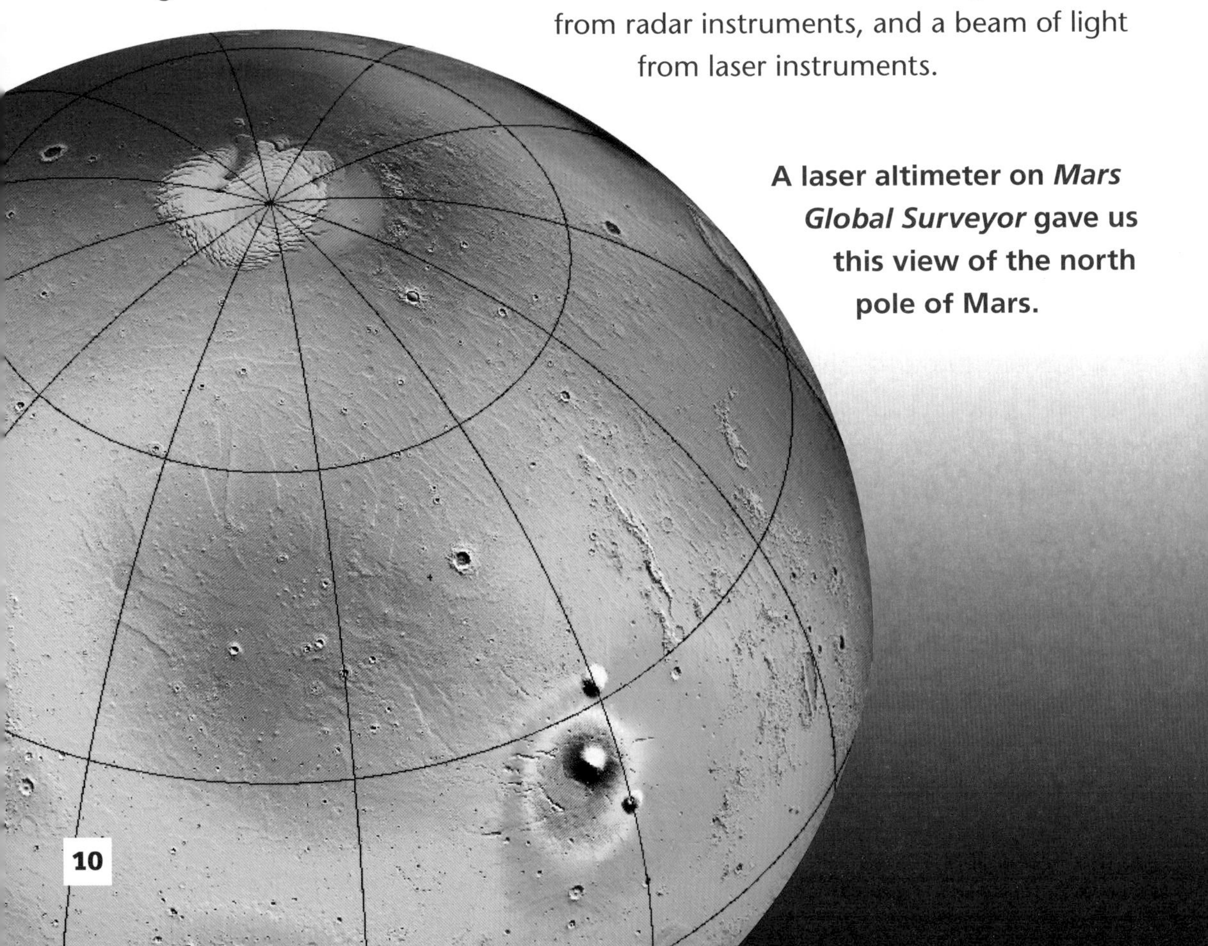

A laser altimeter on *Mars Global Surveyor* gave us this view of the north pole of Mars.

Seismographs, on the other hand, are passive. *Passive* means that these instruments do not send out a signal. The seismograph detects seismic waves.

Another passive instrument is a spectrometer. A *spectrometer* detects and separates different kinds of radiation, much as a prism separates light into different colors. Spectrometers are often made to separate only a certain kind of radiation, such as microwaves.

Spectrometers for Measuring Heat

When objects are warm or hot, they emit a kind of radiation called infrared. Some devices, such as many night-vision goggles, can detect infrared. When you use these goggles, you can see things that you otherwise could not see in the darkness. For example, if you were walking in the woods at night wearing such goggles, you would be able to see animals as bright, glowing creatures. Spectrometers that detect infrared can be used to learn about the heat of a planet and of its atmosphere.

NASA has an infrared instrument called AIRS (Atmospheric InfraRed Sounder) orbiting Earth on a satellite. AIRS measures Earth's ground temperature, air temperature, and also clouds and humidity. Scientists use AIRS to study short-term weather trends and long-term climate changes.

Another type of infrared spectrometer is much farther out in space on the *Cassini* spacecraft, which is studying the planet Saturn. Saturn is a gas giant, made mostly of hydrogen and helium. Far beneath its cloudy atmosphere, Saturn has a layer of liquid helium and hydrogen. Below that, Saturn has a solid core. Although Saturn is far from the sun, it is not completely cold. Saturn makes its own heat, radiating about twice as much heat as it receives from the sun.

The *Cassini* spacecraft has an infrared spectrometer that was designed to measure heat in different parts of Saturn's atmosphere and rings. By doing so, it can map the atmosphere, showing areas where the temperature and pressure vary.

Messenger's spectrometer can measure even the very thin atmosphere of Mercury.

Spectrometers for Studying Atmospheres

Some spectrometers measure differences in molecules to tell them apart. Each kind of molecule absorbs infrared radiation in a slightly different way. Using these special spectrometers, scientists can learn which gases are in a planet's atmosphere.

On July 15, 2004, NASA launched a new instrument called the TES (Tropospheric Emission Spectrometer) on board a satellite. The TES uses infrared to measure the amounts of certain gases in the troposphere. The *troposphere* is the lowest layer of Earth's atmosphere, reaching up to about 10 kilometers (6 miles) high. One of the TES's major goals is to study Earth's ozone. The gas *ozone* is a normal part of the atmosphere. Ozone can be a good thing or a bad thing, depending on where in the atmosphere you find it.

The stratosphere is a layer of atmosphere that lies about 10–50 kilometers (6–30 miles) above the troposphere. The stratosphere contains almost 90 percent of Earth's ozone. Ozone in this part of Earth's atmosphere is called the ozone layer. Ozone in the stratosphere is good for Earth because it filters out most of the sun's harmful ultraviolet light.

However, too much ozone in the lower layers of the atmosphere is harmful to both plants and animals. It reacts with air pollution to form smog. Large amounts of ozone can damage living tissue. The TES is designed to find out more about where ozone in the lowest layer of the atmosphere comes from and how it reacts with other things in the air.

Another type of spectrometer is on board *Cassini,* where it is studying the planet Saturn's atmosphere. A UVIS (UltraViolet Imaging Spectrograph) aboard the spacecraft measures the ultraviolet radiation reflecting off Saturn's rings and atmosphere. In July 2004, *Cassini* found particles of dirt among the ice pieces in Saturn's rings, which may mean that the rings formed when a moon broke apart.

Much closer to the sun, the *Messenger* spacecraft will study Mercury's atmosphere in much the same way. Mercury is a small planet, so its gravity is weak. As a result, Mercury has almost no atmosphere. However, it does have a thin layer of gases. *Messenger*'s spectrometer will measure the infrared and ultraviolet radiation from those gases. This spectrometer can even detect and measure minerals on Mercury's rocky surface.

Io above the surface of Jupiter

Spectrometers for Making Maps

Some spectrometers are designed for mapmaking. They can be used in much the same way that sonar, radar, and laser readings can be used to study the land. However, spectrometers measure not only how the land is shaped but also what the land is made of.

One way scientists learn more about Earth is by using a spectrometer called AVIRIS (Airborne Visible/InfraRed Imaging Spectrometer). Carried onboard special airplanes, AVIRIS measures and maps the plant cover, snow cover, and minerals on the land below. AVIRIS can also measure how moist the air is and the color of the sea.

Far away from Earth, the *Galileo* spacecraft has made maps of Jupiter and its moons using NIMS (Near Infrared Mapping Spectrometer). Jupiter, like Saturn, is a gas giant. NIMS measured the changing temperatures and cloud layers that make up Jupiter's atmosphere. These measurements told scientists that Jupiter has giant thunderstorms. These storms are far greater than any thunderstorms that occur on Earth.

NIMS also measured minerals and temperatures on Jupiter's moons. One of Jupiter's moons is named Io (EYE•oh). Io is not much bigger than Earth's moon. On Io there are many active volcanoes. Scientists learned from measurements taken by the *Galileo* spacecraft that the volcanoes on Io are hotter than the volcanoes on Earth.

The *Cassini* spacecraft also has a mapping spectrometer. Remember Titan, the moon with the atmosphere? *Cassini* has used its mapping spectrometer to take pictures right through Titan's dense atmosphere. These pictures show a land with surface features: lines and circular features that are different from anything on Earth.

The mapping spectrometer can also measure the chemicals and minerals on Titan's surface. Because the spectrometer can collect these measurements, scientists are putting together maps of what the surface of Titan is made of.

Beyond the Gas Giants

From Earth, through even the most powerful telescopes, Pluto looks like a whitish spot. Some scientists argue that Pluto, the ninth planet, is not really a planet.

In the last few years, several objects nearly the size of Pluto and with a similar composition have been spotted in the far reaches of the solar system that stretch beyond the gas giants. That part of the universe is called the Kuiper Belt.

A probe called *New Horizons* is scheduled to be launched from Earth, in 2006, and will reach faraway Pluto in 2015. The probe will examine the tiny, icy planet and its moon, Charon. The probe will also look for other large objects nearby in the Kuiper Belt. Called Kuiper Belt Objects (KBOs), these objects are made of ice and rock, just as Pluto is. Before the 1990s, no one even knew that KBOs existed. Now, with better Earth-based telescopes and planet-measuring devices, scientists are beginning to think that Pluto itself is a KBO, and not a planet.

As it passes by Pluto, *New Horizons* will measure ultraviolet emission from Pluto's atmosphere. It will use its instruments to make a detailed map of both Pluto's and Charon's surfaces. It will also analyze surface composition and measure temperatures.

At its closest approach to Pluto, *New Horizons* will take visible-light pictures and near-infrared pictures. Those pictures should reveal features on Pluto as small as 60 meters (200 feet) across. After the probe passes Pluto, it will look back at the dark side of the planet and check for an atmosphere or the presence of rings. From there, *New Horizons* goes on to explore the Kuiper Belt, using its instruments to find a KBO.

KBOs, including possibly Pluto, are thought to be leftover materials from the formation of the solar system. By measuring any that it finds, *New Horizons* will be able to tell scientists more about what the solar system looked like before the nine planets formed.

Or is that eight planets? In 2006, the International Astronomical Union defined a planet as a body that orbits the sun, is roughly spherical in shape, and is massive enough to clear its orbit. They reclassified Pluto as a "dwarf planet" because it does not have enough mass to clear its orbit.